The Bouldon Furnace
Shropshire

Bernard O'Connor

Bernard O'Connor

Copyright Bernard O'Connor 2019

ISBN: 978-1-71662-773.6

The Bouldon Furnace

Researching Bouldon's history I found John Roque's 1752 map of Shropshire on which 'Bouldon Furnace' was marked. It shows two rectangular buildings on the south side of what is known today as Pye Brook on a lane or track running west from 'Clee Town' and then southwest to meet a lane going south to 'Hopton on the Hole'. There was no lane going down the valley towards 'Peeton' and no other buildings marked. Interestingly, the 1753 map had 'Iron Mines' and 'Coal Pits' marked to the east on Abdon Burf and Brown Clee Hill.

The furnace was not marked on Robert Baugh's 1808 map or on Greenwood's 1828 map but there was a 'Paper Mill', more buildings and lanes recognisable today. However, the 1842 Tithe and Appointment map of Holdgate provided the names of the fields. It shows 'Furnace Wood' and 'Furnace Field' on the western slopes of Pye Brook.

An 1844 map found in the Victoria Country History (VCH)'s account of Bouldon also shows these fields, the paper mill, another mill beside Pye Brook, Bank Farm and Bouldon Farm.

The first 25-inch Ordnance Survey maps, drawn in the 1880s or 1890s, no longer show the paper mill but a cornmill, known today as Bouldon Mill. It also shows a leat, or mill race, a channel that led from where a weir across the brook. The water would have flowed down the leat was used ro drive a waterwheel at the furnace.

The exact location of Bouldon furnace shown on the 1728 map is not easily identifiable today but the most recent Ordnance Survey map would suggest it was on the site of or near to Bank Farm.

In the Victoria County History's account of Bouldon's economic history, it referred to an iron furnace in the 1640s. 17th century iron making required iron ore and charcoal. Charcoal was burnt to provide the heat to melt the iron ore. When it was discovered that adding limestone to the furnace acted as a flux to help separate the waste rock from the iron is unknown. As shall be seen, iron ore was mined on the western slopes of the Clee Hills about five kilometres to the east-northeast, Abundant wood for charcoal was also available from coppices around the Clee Hills and limestone deposits were found on both sides of the valley of Pye Brook.

The first Geological Survey map of Corvedale shows two outcrops of Psammosteus Limestone between the lower Downton and upper Ditton Old Red Sandstone. Today, the strata has been renamed because the Victorian geologist who gave it its name wrongly classified fossils found in the limestone as psammosteus anglicus, a jawless fish similar to a lamprey. Modern geologists identified it as traquairaspis symondsi. As the same geological strata was noticed at Bishop's Frome, between Hereford and Worcester, it has been renamed Bishop's From limestone.

There are disused limestone quarry workings less than 100 metres from Bank Farm, just above the 175-metre contour line. Access to the quarry is via an overgrown track running southeast from Bank Farm, not the lane running east. Once past the metal field gate, a sunken track winds south and then southwest up the side of the hill. Centuries of horses, cattle and sheep's hooves and the ruts of cartwheels cut into the soil which was then washed down the track during heavy rain.

On the left-hand side of the track there are extensive views north and east overlooking the valley of Pye Brook towards Heath On the right-hand side there is the limestone quarry with steep banks and a flat open floor (SO548848) where one imagines workshops and associated buildings were located.

Whether the original furnace was located at the quarry site is uncertain as there would have been a need for a water supply to cool down the metal. Whilst it is possible a well was sunk, buckets of water from the brook could have been carried to the quarry and emptied into wooden barrels or tubs. A line of disused limestone quarries follow the line of the strata on the northern slopes of Pye Brook.

What agreement the landowner had with the owner of the furnace and what financial agreement they had is unknown, as is how many labourers were employed, whether they lived in accommodation on the site or walked to work, what hours they worked, what wages they received, what food they ate and what clothes they wore.

Smelting ore began in prehistoric times and manufacturing iron was practiced before the Roman period. Charcoal-fired furnaces had been introduced into Britain from Belgium by the end of the fifteenth century. Research by historian Trevor Rowley showed that in 1540, the monks of Wenlock Priory had ironstone quarries in Shirlett Forest, southeast of Much Wenlock, which supplied two iron foundries, and monks at Buildwas Priory had a forge, where iron bars could be re-heated and hammered into a variety

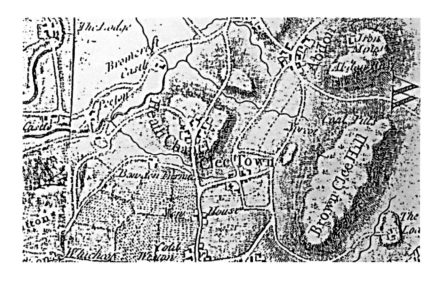

Extracts from John Roque's 1752 map (Shropshire Archives)

Extract from Robert Baugh's 1808 map (Shropshire Archives)

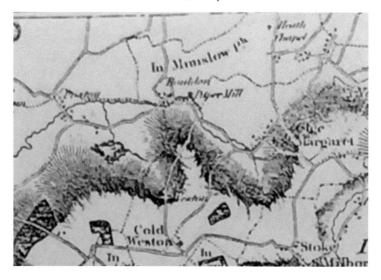

Extract from Greenwood's 1828 map (Shropshire Archives)

The Bouldon Furnace

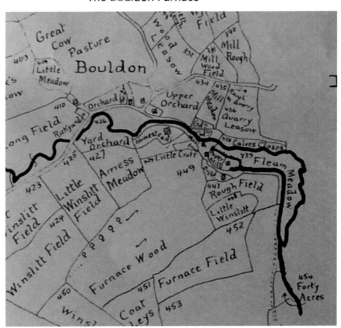

Extract from the 1842 Tithe and Appointment map
(Shropshire Archives)

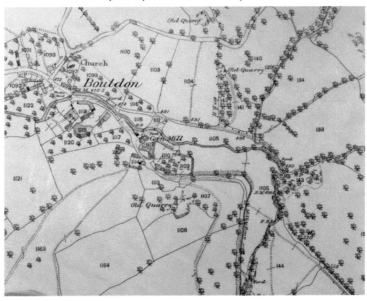

Extract from 1880s or 1890s 25-inch OS map showing part of the
leat (Shropshire Archives)

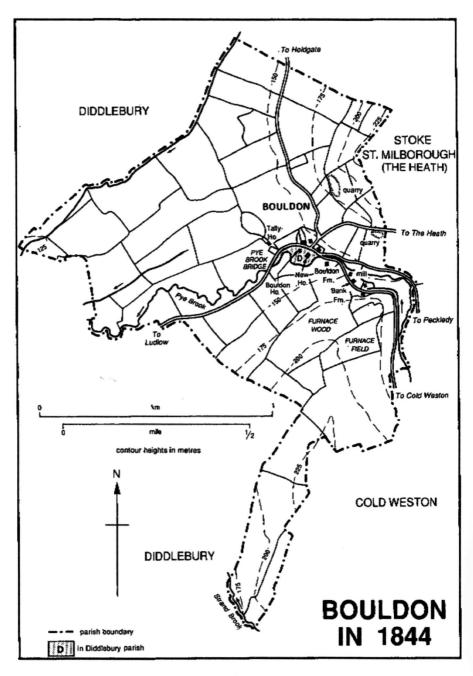

(https://www.british-history.ac.uk/vch/salop/vol10/pp147-151)

The Bouldon Furnace

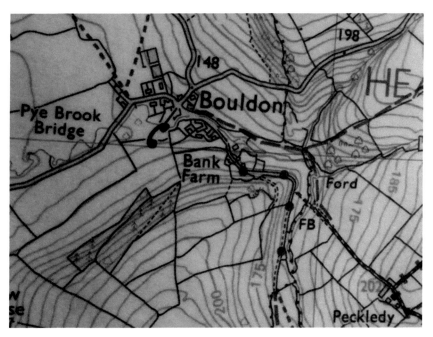

Extract from 1:25000 OS map

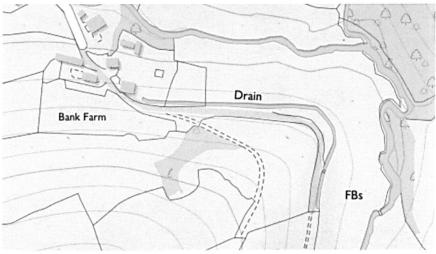

Extract from OS map showing Bank Farm (furthest east of the buildings), the track to the quarry and part of the leat, marked as a drain.

Georg Agricola: De re metallica (1556) https://www.gildersome.net/iron-working.html

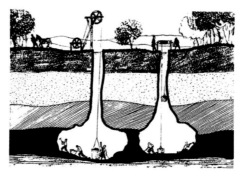

Bell pit mining was used to extract ironstone from Monkey's Fold on the Brown Clee (https://www.blythtown.net/articles/early-coal-mines-126)

of shapes. (Rowley, Trevor, *The Shropshire Landscape*, Hodder and Stoughton, 1972; Rowley, Trevor, *The Landscape of the Welsh Marches*, Michael Joseph, 1986, p.228)

King Henry VIII's dissolution of the monasteries between 1536 and 1541 led to an increase in private land ownership and opportunities for the new landowners to develop the economic potential of extracting any available mineral deposits from their land. (Rotherham, Ian D. Jones, Melvyn and Handley, Christine (eds.), 'Working and Walking in the Footsteps of Ghosts'. Vol. 1: *The Wooded Landscape*, Wildtrack Publishing, 2003/2012)

Rowley reported that in 1561, 'licence was granted to Sir William Acton to fell trees necessary for making iron and to tell timber from Shirlett for his "lately buylded and set up iron mills in Morveld (Morville)." Because of the shortage of timber Crown authority was required to clear woodland within fourteen miles [22.4km] of the River Severn at this time. Supplies of accessible timber were badly needed for the emerging iron and glass industries and these were largely supplied by planting coppices. With the widespread creation of iron furnaces between Willey Park to the east and Bringewood in the west, the first attempts at controlling coppicing were made. [...] Much of Shropshire's surviving woodland was originally planted for the iron industry, and it is interesting to observe that some of the most densely wooded parts of the county,,, owe their tree cover to the needs of the charcoal blast furnace.' (Rowley, op.cit. pp.217-8)

Research by Richard Hayman showed that the blast furnace originated in the early medieval period but did not arrive in Britain until the late 15th century. The first was constructed on the site of a bloomery by French workmen at Newbridge in the Weald, southeast England, in 1496. (Hayman, Richard, The Shropshire Wrought-Iron Industry c1600-1900: A study of technological change, DPhil thesis, Birmingham University, 2003, p.22)

Research by Dr David Poyner into the iron workings in the Highley area, about 10km south of Bridgnorth, revealed that by 1576, Robert Dudley, the 1st Earl of Leicester, had two forges constructed on the River Rea near Cleobury Mortimer (c. 1563 SJ611055 and c. 1576 SO71137642). As Dudley had family connections with the ironmasters on the Weald, he recognised the profits to be made from owning the rights to the raw materials.

Under the management of John Weston, charcoal was obtained from the Wyre Forest, iron ore from Billingsley, Chorley, Neen Savage and the Clee Hills, and water from Dowles Brook. In the 1580s, Dudley was reported to have made around £1,600 per year from his ironworks. (http://www.highley.org.uk/page28.html)

In Richard Hayman's 2003 thesis on Shropshire's wrought'iron industry, he stated that,

> In South Shropshire, the earliest blast furnaces were erected at Cleobury Mortimer on land granted in 1563 to Robert Dudley (1532-88), later Earl of Leicester. Two blast furnaces, subsequently known as Cleobury Park and Furnace Mill, had been built by 1576. Both were formerly bloomeries, referred to later A finery forge [where cast iron was re-heated and forged into wrought iron] was let to Stephen Hadnall in 1571, and another to John Weston in 1576, both on the River Rea and subsequently known as the upper and lower forges. A third forge was built in 1597 by Richard Cook at Boraston on the River Teme, on the border with Worcestershire and less than a mile north east of Tenbury Wells. (KWG Goodman, Hammerman's Hill: The land, people and industry of the Titterstone Clee Hill area of Shropshire from the sixteenth to the eighteenth centuries, PhD thesis, University of Keele (1978), pp 97-8) Cleobury Mortimer manor was purchased by Rowland Lacon soon after 1608, while the adjoining manor of Barnsland to the east was purchased by George Blount in 1601. The Lacon and Blount families, related by marriage, dominated the South Shropshire iron industry in the early seventeenth century. The Cleobury Mortimer furnaces and forges soon came under the direction of the Blount family, although the blast furnaces ceased operation and the last reference to them in parish registers is in 1633. Boraston Forge also ceased work in the seventeenth century. (Ibid., pp 98, 106; M Baldwin, 'Ironworking in Cleobury Mortimer', Cleobury Chronicles, 3

Sketch of an early bloomery
http://shropshirehistory.com/iron/iron_making.htm

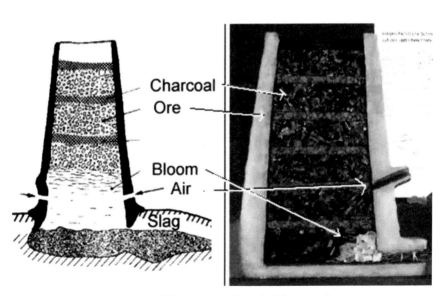

Diagrams of early bloomeries
(https://www.tf.uni-kiel.de/matwis/amat/iss/kap_a/illustr/ia_2_4.html)

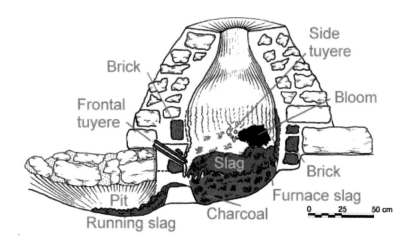

Permanent bloomery, Archäologie Baselland; Swiss (https://www.tf.uni-kiel.de/matwis/amat/iss/kap_a/illustr/ia_2_4.html)

Slag, possibly from a bloomery, found in a field wall by Pye Brook upstream of Bank Farm. (Courtesy of Bernard O'Connor, July 2019)

(1994), pp 40-3) The cessation of smelting in Cleobury Mortimer necessitated acquisition of other blast furnaces further afield in order to supply its forges. In 1623-4 Francis Walker, a clerk at Cleobury Mortimer, leased the blast furnace and forge at Bringewood on the River Teme in Herefordshire (formerly Shropshire), which stayed in the family until 1698 (Bayliss, D.G. 'The Effect of Bringewood Forge and Furnace on the landscape of part of northern Herefordshire to the end of the seventeenth century', Transactions of the Woolhope Naturalists Field Club, 45/3 (1987), p 722; Hayman, op.cit.)

In the 1590s, there was a charcoal iron works on the River Teme at Bringewood (SO45457498 and SO45407496) in Herefordshire, about 6km west of Ludlow. Walter Devereux, the first Earl of Essex, was reported to have ordered its construction and it is thought to have been built for Robert Dudley after the manors of Wigmore and Burrington and Bringewood and Mocktree Forests had been granted to his nominees, Gelly Merrick and Henry Lyndley. (King, op.cit. TNA LR1/136, ff244v-248v)

King suggested that the iron ore might have been mined locally but local tradition has it that packhorses brought sacks of iron ore down from the Clee Hills across Dinham Bridge in Ludlow, up the 'donkey steps' in Mortimer's Forest, over Bringewood Chase to the furnace. Coppiced wood for charking was available on the nearby hills and limestone was quarried nearby.

When the land reverted to the Crown, the Bringewood iron works were let to Sir Henry Wallop who sublet them in 1619 to local ironmaster, Edward Vaughan. By 1623 they were being run by Francis Walker whose descendants operated it until 1695 when Job Walker went bankrupt. Why his business failed is unknown. (https://en.wikipedia.org/wiki/Bringewood_Ironworks)

Thomas Southcliffe-Ashton reported that iron smelting first took place in Shropshire near the Wrekin, where furnaces were in operation in about 1618 at Willey, (SO 672980), about 6km east-southeast of Much Wenlock and in about 1630 at Leighton (SJ611055), about 6km north of Much Wenlock. (Southcliffe Ashton, Thomas, *Iron and Steel in the Industrial*

Revolution, Manchester University Press, (1953) and 1951, p.18) Sir Basil Brooke, Lord of Madeley Manor, built the old blast furnace at Coalbrookdale in 1638. (Rowley, op.cit. p.217).

Rotherham, Jones and Handley reported that by the seventeenth century there were five furnaces between Willey and Wombridge, east and northeast of Much Wenlock, with more to the south. (Rotherham, Jones and Handley, op.cit.)

By 1630 there was a furnace operating at Catherton, about 6km northwest of Neen Savage (SO637777). In 1635, Sir Richard Newport constructed a blast furnace at Ercall Magna, northeast of Shrewsbury. At that time or shortly afterwards, there was a forge at Sheinton, about 6km north of Much Wenlock. By 1638, William Boycott and William Fownes leased the Newport works. In 1641 there was a furnace at Ditton Mill, near Hopton Wafers. In 1654, Sir Humphrey Brigges of Ernestrey [Earnstrey] Park, about four kms northeast of Bouldon, had the 'liberty of getting and carrying away all mines of ironstone off the Brown Clee Hill.' As Lord of Abdon Manor, it is thought he arranged for the furnace on Tug Brook to be contructed. (http://www.highley.org.uk/page28.html; Hayman, op.cit.)

When the Bouldon furnace was first built is uncertain. Peter King, who researched the British iron trade, suggested it was about 1630, Rowley gives the date as 1644. The earliest documentary evidence shows it was in operation in the early 1640s during the English Civil War. (King, Peter, *A Gazeteer of the British Iron Industry, 1490—1815, Vol. 1 & II*, British Archaeological Reports, Oxford, 2019; www.ehs.org.uk/ dotAsset/6a439da7-0919-41de-acaf-b37ed608496e.xls)

Barrie Trinder reported that Bouldon's 17th century furnace was amongst the earliest in Shropshire, with others at Abdon (SO 567867), Coalbrookdale, Ifton Rhys, Leighton, Tilsop and Wombridge. King estimated the Abdon furnace to have been constructed by 1654. Another charcoal blast furnace that used Brown Clee ironstone were at Cinder Hill, Charlcotte (SO SO63888608).(King, op.cit; Trinder, Barrie, *The Industrial Archaeology of Shropshire*, Phillimore, 1996, p.30)

Ironstone or iron ore was located by prospectors walking up river valleys and stream beds, searching for signs of red, brown or orange coloured rocks which was indicative of quantity of ferrous oxide. Nodules of ironstone were found between the

Portrait of Baron Mountjoy Blount (c.1597 – 1666) by Sir Anthony van Dyck. He was Charles I's Master of Ordnance in 1634. (https://en.wikipedia.org/wiki/Mountjoy_Blount,_1st_Earl_of_Newport)

Carboniferous coal measures which had been exposed by glacial erosion but subsequently by river erosion.

    The nearest coal and iron ore deposits to Bouldon were on the upper slopes of Abdon Burf and Clee Burf. Once the iron ore or ironstone strata were located, bell pit mining was used. The overlying earth was removed to expose the seam and the iron ore nodules dug out and brought to the surface using a

Diagrams of undershot and overshot water-powered bellows used in a blast furnace (Top: http://explorepahistory.com/displayimage.php?imgId=1-2-FBF Courtesy of Courtney Howell Bottom http://1.bp.blogspot.com/-c5J1e0oxBRA/VDtrdsD16eI/AAAAAAAACdo/fhuVDn2Q5V8/s1600/GermanForge.JPG)

The Bouldon Furnace

Section showing the water-powered bellows used to blast air into the furnace. (http://www.thepotteries.org/shelton/150/5.htm)

18th century blast furnace (https://c8.alamy.com/comp/2C161F8/18th-century-illustration-of-a-outside-view-of-a-blast-furnace-published-in-a-diderot-pictorial-encyclopedia-of-trades-and-industry-manufacturing-a-2C161F8.jpg)

View of a blast furnace showing the block and tackle lifting apparatus (https://www.tripadvisor.co.uk/LocationPhotoDirectLink-g60854-d294403-i282687989-Hopewell_Furnace_National_Historic_Site-Elverson_Pennsylvania.html)

PUDDLING.

Engraving of a puddling furnace. (https://images.app.goo.gl/dHpK4pvQyRLwjNHz8)

Pouring molten iron into a casting box (https://www.buildingconservation.com/articles/orncastiron/)

Ingots of pig iron which would have been sold to forgers and blacksmiths (https://staffordcountymuseum.com/artifact/pig-iron-ingots/)

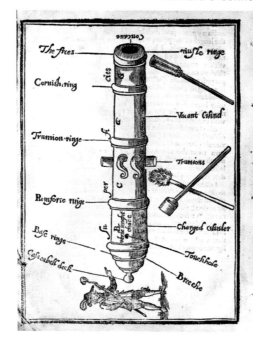

(Roberts, John, *The Compleat Cannoniere*, London, 1652, pp.24-5)

Model of a Civil War cannon (https://www.ebay.co.uk/p/King-and-Country-English-Civil-War-Cannon-PNM036-Pnm36/1139307020)

Photographs of a cannon ball found in the garden of Jacks Barn, Bouldon. (Courtesy of John and Jill Perks)

winch. Broken into manageable pieces it was sacked up ready to be transported to the furnace by pack horses or donkeys. Before the sides of the pit collapsed, another pit was sunk nearby with the waste soil and rock thrown into the adjoining abandoned pit, leaving a landscape pitted with depressions.

The place where the smelting took place was originally called a 'bloomery'. This was a hand-made furnace constructed from a circle of broken stones held together with mud, about a metre tall. It had a hole in the base to insert kindling and logs to start a fire. Charcoal was added through the opening at the top, followed by the iron ore. A sketch and diagrams of early bloomeries can be seen in the illustrations.

As iron ore has to be heated to 1,200° Centigrade for the metal to liquify, hand or foot powered bellows were needed to blow air into the fire. However, early furnaces were never at a hot enough temperature to completely melt the iron. Instead a spongy mass known as a 'bloom' was left at the base of the fire. This was removed with iron tongs and beaten with a hammer on an anvil into a bar which forced out any remaining stone. Although it produced iron, it was poor quality. It was brittle, would often crack when hit, and rusted in wet weather. This was because the metal contained impurities like waste rock and charcoal ash. Bloomeries produced about 30 tons of iron a year.

When the technique of adding limestone to the furnace was introduced, the melted limestone acted as a flux. It mixed with the ironstone leaving the molten iron to run free, a process which produced much better quality iron,

When the fire was out, the hearth was emptied of the ash and waste rock, known as slag or clinker. Shovelled into wheelbarrows or carts, it was taken away to be emptied down the side of the valley to become a slag heap. The Bouldon slag consists of black, grey or bottle-green glassy rock. The slope down towards Pye Brook by Bank Farm (SO548849), Jacks Barn (SO547849) and Pool Cottage (SO547849) is mostly black soil from burnt charcoal and contains much slag. Pieces of slag can also be found in the soil in the gardens of properties lower down the valley, thought to have been washed downstream during flood conditions. Slag was also used to fill in ruts in local tracks and muddy depressions by field gates caused by sheep and cattle hooves and cartwheels. It can be seen today, especially in wet weather when shiny black stones are visisible.

As limestone was heavier to transport than iron ore and cordwood, the bloomeries were located close to the quarry. Once the overlying soil was removed to expose the limestone, crowbars, pickaxes and perhaps gunpowder was used to break off blocks of the rock. Whether these blocks were then burnt in a fire and hammered into smaller pieces before being added to the bloomery is unknown.

Beside the bloomery there would have been a 'casting shed' with a bed of sand, up to a foot thick, which had a mould dug out. This was a trench, maybe six feet in length and six inches deep, with smaller trenches, maybe a foot long, at right angles. It looked rather like a sow with sucking piglets. The molten iron was 'puddled' using a 'puddling iron', to push it into each trench where it was allowed to cool and solidify. Depending on what was required, other moulds were made for the iron to be poured into, for example cannon and fire-backs, the iron plate at the back of the fire to protect the wall,

Care had to be taken to avoid splashing so it was likely that the workers wore wooden clogs as leather would burn if they stood on splashed iron. When cold, the sand could be brushed away and the iron hammered to break it into transportable pieces. These pieces, known as 'piggs' or 'pig iron', generally contained 3-4% carbon which made them too brittle to be malleable so needed to be further refined. Loaded onto carts, they were taken to be sold to the 'forgers' or 'blacksmiths'. (Transactions of the Shropshire Archaeological Society (T.S.A.S.) 4th ser. Vol. xi. p.109; Harman, Richard, The Shropshire Wrought-Iron Industry c1600-1900: A study of technological change, DPhil thesis, Birmingham University, 2003, p.19)

Charcoal was made from the slow burning of coppiced wood. Large-scale coppicing was introduced in the sixteenth century to supply ironmasters. Trees of at least

Sunken track leading up to limestone quarry and the probable site of the original charcoal furnace Bernard O'Connor)

The Bouldon Furnace

View from quarry, first suggested site of furnace, to Bank Farm, the second site of furnace, and the track leading to Bank Farm (Bernard O'Connor)

Slag heap below Bank Farm and Pool Cottage showing the black soil from the burnt charcoal and glassy lumps of slag. (Bernard O'Connor)

Weir on Pye Brook. Note the stonework on either side. Flo[ods] have probably destroyed much of the upper section. (Bern[ard] O'Connor)

Stone-lined leat leading from the weir along the contou[r] to the furnace at Bank Farm (Bernard O'Connor)

The Bouldon Furnace

The leat, now covered, fed to an eight feet deep trench along the south side of Bank Farm which turned a waterwheel. (Bernard O'Connor)

The axle of the waterwheel, now gone, would probably have been where the lower left-hand window is. Note the newer stonework (Bernard O'Connor)

Sketch of mill pond, sluice gate, waterwheel, bellows and furnace at Cleobury Park.

A cut-away sketch of a water-powered blast furnace.
(http://www.highley.org.uk/page28.html)

Foot-powered bellows (https://www.bbc.co.uk/
northernireland/yourplaceandmine/topics/myth/
A762167.shtml)

Water=powered bellows at Sugus Iron Mill (https://
commons.wikimedia.org/wiki/File:Saugus_Iron_Mill_-
_forge_with_bellows.JPG

18th century finery from Diderot's Encyclopédie

18th century forge hammer from Diderot's Encyclopédie

20 years old were chosen. The trunk was cut at ground level and the new straight branches were cut off. Hazel and hornbeam could be cut after seven years but alder, ash, beech and birch could be cut after three years' growth. The branches were cut into pieces and sold to 'colliers', charcoal burners, for 'charking'. The cut branches were stored in piles known as a cord, 1,22m. wide, 1.22m high and 2.44m. long. One cord weighed about one ton. There are documents stating that cordwood was brought to Bouldon which, if true, suggests the process of 'charking' was done by the 'bloomery'. However, as three cords of wood were needed to produce one cord of charcoal, it is possible the charking was done in the coppice.

Research by Jon Van Laun into Herefordshire's charcoal iron workings showed that about 3¼ acres of trees needed to be coppiced to produce the charcoal needed to produce one ton of iron. 3¼ acres is about one and a half football grounds. With furnaces across Shropshire and Herefordshire and thousands of tons of iron being produced each year, it contributed to the widespread deforestation. It is very likely that Wynett Coppice on the west-facing slope north of Bouldon was cut to supply the furnace.

To make charcoal, cut branches were piled up, covered with earth with an opening at the base to set the wood alight. This allowed the wood to smoulder rather than burn with a flame. After several days, when the fire was out, the earth was removed and the 'chark', burnt wood or charcoal, was collected for sale to the iron master in charge of the 'bloomery'. Crushed in a bucket with a log, the powdered charcoal was emptied on top of the burning wood, followed by the iron ore and limestone. (Ibid.)

Pack horses, or donkeys, brought sacks of iron ore down the track from the bell pits on Monkey's Fold, the west-facing slopes of Clee Burf, through Cockshutford and Clee St Margaret, past 'The Boot' (previously a public house), and over the hill past Peckledy Farm, down the sunken track to Pye Brook and along the northern side of the brook to a ford near what later became Bouldon Mill (SO547850). From there the packhorses would have been led up the lane to the Bank Farm site. Bundles of cordwood cut from coppices along the lower slopes of the Clee Hills was also brought in for charking.

The Victoria County History provided details of the furnace in its entry on Bouldon's economic history. 'An iron furnace at Bouldon was said to have been established by one of the Smiths, lords of the manor.' The reference was to an undated 18th century letter written by Samuel Fowler to William Mytton, in which he reported that 'At Bouldon there is an ironstone Furnace, which melts the best Iron Pigs (as is said in England). The Iron Stone they have in great plenty from the Brown Clee Hill about two miles distant.' He added that the furnace was first built by Charles Carrington Esquire, one of the Smiths of Aston Hall, a 16th century Manor House (SO 509866) in Munslow Parish.

The local Lord of the Manor in the 18th century was Sir Charles Smyth of Burford (1598-1665) who was appointed the 1st Viscount Carrington of Connaught, Ireland, in 1643. (Shropshire Archive X7381/112/753A, IMG1037; https://www.british-history.ac.uk/vch/salop/vol10/pp151-167; https://en.wikipedia.org/wiki/Viscount_Carrington)

In 1647 the furnace was in the possession of Sir William Blount, presumably as tenant. (Rowley, R.T. 'History of South Salop. Landscape', Oxford University, B.Litt. thesis, 1967, p.170; Cokayne, George Edward, The Complete Peerage of England, Vol. ii. pp. 202–3.) Sir William Blount was likely related to Baron Mountjoy Blount (c.1597 – 1666), the first Earl of Newport, who was King Charles I's master of ordnance in 1634. The Blount and Lacon families, related by marriage, dominated the South Shropshire iron industry in the early 17th century with forges at Cleobury Park (SO 711764) and Furnace Mill, Boraston. (Hayman, Richard, *The Shropshire wrought Iron industry between c.1600 and 1900: A study of technological change,* University of Birmingham, July, 2003)

The English Civil War between King Charles I's Royalist army and Oliver Cromwell's Parliamentary forces started in 1642. According to the Wikipedia page on the Civil War in Shropshire,

> Politically, the English county of Shropshire was predominantly Royalist at the start of the civil war. Of the county's twelve Members at the Long Parliament called in 1640, eight would fight on the Royalist side and four for Parliament. Control of the area was important to the King as Shropshire was a gateway to predominantly

Royalist Wales as well as to keep in contact with the north-western counties and the western port links with Ireland. Parliamentary control of Shropshire was achieved after the capture of its last Royalist garrison by Parliament in 1646.

The week after raising his standard at Nottingham, Charles I proceeded into Shropshire, arriving in Wellington on 19 September. On 20 September he issued the Wellington Declaration promising to preserve the Protestant religion, laws, and liberties of his subjects, and the privileges of Parliament, and inspected his troops below the Wrekin. From Wellington he marched to Shrewsbury, where he was joined by his two sons, [Charles] the Prince of Wales and James, Prince Rupert, and great numbers of noblemen and gentlemen, and established a mint in the town. He remained there until 12 October, when he marched to Bridgnorth, and from there advanced to Edge Hill, in Warwickshire, where the first pitched battle of the First Civil War was fought. (https://en.wikipedia.org/wiki/Shropshire_in_the_English_Civil_War)

When the conflict started, the Royal Artillery had an urgent need for more cannons and cannon balls. In 1643, Francis Walker was recorded as 'clerk of the furnace' at Bouldon, a position he held until at least 1661. He was related to Richard Cressett (d. 1677) of Upton, one of the wealthiest men in Shropshire. (Shropshire Archives 1359, box 20, Thomas Martin to Richard Holman, 26 June 1661 (copy); Goodman, Ken, 'Hammerman's Hill: Land, People, and Industries of Titterstone Clee Area 16th to 18th Centuries.' (Keele University, Ph.D. thesis, 1978), p.106)

Mr Thomas Fisher was reported to have raised a company in Ludlow in the Civil Wars for the king., and took ordnance from Bringewood Forge to defend the town. (Trans. Shrops. Arch. Soc. (3rd Ser.), q.v.p.115)

It seems likely that some of Walker's Bringewood 'mettlemen' or 'forgemen' were brought over to work at Bouldon. Walker was reported to have been paid (the sum was not specified) for making 63 tons of ordnance for the king, delivered from Bouldon to Bridgnorth and Shrewsbury. As two horses were needed to pull a cart carrying three tons, there would have been many journeys to supply the Royalist troops.

Most military cannons were supplied from furnaces in the Weald, southeast England. Stephen Bull, the curator of Military History and Archaeology at Lancashire Museums, reported that 'Cannon certainly were produced for the Royalists in Shropshire, but the numbers were undoubtedly relatively small. An order for 44 guns from iron master Francis Walker's Bouldon works near Ludlow for the defence of Bristol was probably completed during 1643, a request for more followed later.' (Bull, Stephens, *The Furie of the Ordnance Artillery in the English Civil Wars,* Boydell Press, 2008, p.75)

In 1644, Walker made a similar quantity, including a gun for the defence of Ludlow. (Calendar of State Papers, Domestic, 1641–3, p.488; 1644, p.108; V.C.H. Salop. Vol. i. p.473.) Hayman reported that there was 'evidence for the manufacture of four guns which were destined for a ship at Chester, and also for a supply of shot – but nothing from which a train of modern field artillery could have been created. An interesting piece of circumstantial evidence as to what the Royalist command might have construed as Shropshire's most significant munitions product emerged in 1664. For in that year Charles II granted the Boycot family a coat or arms upon which were no cannons but three granados.' (Hayman, op.cit.)

It was presumed that Walker worked the furnace under licence from the Blounts and was supplied with cordwood from Ditton Priors and elsewhere, ironstone from Brown Clee, and the local Bouldon limestone. (Salopian Journal, 5 Aug. 1795; Birmingham University Library, Mytton Papers, iv. p.753)

In 1654, Richard Walker, the Bringewood ironmaster, leased Down Furnace, near Cleobury Mortimer (SO636748) from Sebastian Legas and in 1690 the coal and limestone of Legas's Craven estate passed to Job Walker of Bringewood, who went bankrupt in about 1695. Why he went bankrupt was not noted but it could have been a combination of depletion of iron ore, expense of charcoal, introduction of coke ovens and competition from other ironmasters. (http://www.shropshirehistory.org.uk/html/search/verb/GetRecord/theme:20091104222852)

On 18 January 1690/1 Sir Humphrey Brigges, who had acquired the Haughton estate near Morville through marriage, agreed to lease to Thomas Lowbridge and James Wilmott of Hartlebury, near Kidderminster, iron mines on the Clee Hill at 2s per dozen for every strike of iron. It is probable that they supplied the Bouldon furnace. (Shropshire Archives X5735/2/2/2 Brooke papers)

Another article reported that 'The Chides of Kinlet, members of the [Walker] family, were interested in the iron trade from 1670 to 1730, being connected with the little furnaces at Charlcott, near Bridgnorth, and at Bouldon, or 'Bowdon's furnace' near Ludlow, which in 1644 had been ordered to supply 'a piece of nine-pounder bullet' [4.08kg] to assist in the defence of Ludlow. It is probable that the furnace at Bouldon was connected with the distant forge at Bringewood, of which members of the Knight family were owners or tenants.' (Trans. Shrops. Arch. Soc. li, p.473)

There would have been a 'casting shed' beside the Bouldon furnace where cannons was made. In P.J. Browne's article on cast cannon, he reported that cannon were cast in a loam mould, a similar process used in casting church bells. Building the mould involved a spindle with a strickle board, a piece of wood with the shape of the cannon cut out. Around the spindle was wound a grass rope, covered in a friable material to form the shape of the cannon. The strickle board was dragged across the surface to make the exterior round. The mould was made longer than the gun, with a gunhead beyond the muzzle. To this were attached models of the trunnions and decoration. This was then covered in loam, which is a mixture including a clayey material, dung, and straw.

After being baked, fired in the furnace, the grass rope was pulled out with the loose material around it. The resultant mould was then stood upright in a casting pit in front of the furnace mouth and a core to mimic the interior of the gun was mounted in it. Iron was then run from the furnace into the mould to fill it including the gunhead. The objective was to provide somewhere for slag to float into so that the cannon itself would be free of slag.

It usually took six hours for the iron to cool down and harden. After casting, the mould was broken off and the core removed, and the gunhead cut off. After this the bore needed to be reemed out to remove any blemishes, using a boring mill. The outside was 'fettled' to remove any uneven surfaces. A block and tackle of rope and chains would have been used to move the cannon around the casting shed and eventually onto a cart.

There was a temptation to use iron to repair holes in the casting, bubbles called honeycombs, but by the 18th century the Board of Ordnance condemned any guns where this practice was detected as the metal used for the repair was liable to fly off when the gun was fired with lethal results for anyone stood nearby. Mortar cannon were made in the same way but they were wider and shorter than the traditional cannon. Whether mortar cannon were made at Bouldon is unknown. (Browne, P.J. 'Cast cannon: the 15th to 19th century', *Foundry Trade Journal*, 108 (1960), pp.163-5; Email communication with Peter King, 22 August 2020)

The technology needed to drill out the bore of the cannon was not developed until the 1770s, by which time Bouldon was producing pig iron. Some cannons were so accurate they could land a 32-pound (14.5kg) cannon ball on a target three miles away. Youtube has videos showing how people today can make cannons. (https://www.youtube.com/watch?v=e60CAAAhshI; https://www.youtube.com/watch?v=NNMF6BDL8nk)

According to the historyonthe net website,

> The cannons used in the Civil War were very heavy and difficult to move. The largest needed a team of 16 horses to move them. More commonly, smaller cannon were used but even these required at least 4 men to move them. For this reason they had to be put into position before a battle began. The missiles fired from the cannon were usually balls of iron, but sometimes stones were used. After the cannon had been fired the soldiers operating it had to go through a strict procedure of cleaning, loading the weapon and loading the gunpowder before it could be fired again. Aiming was difficult and the cannon were more effective as a means of instilling fear into the enemy

than actually causing damage.' (https://www.historyonthenet.com/english-civil-war-weapons)

To make a cannon ball or round shot, hemispherical moulds, linked together like a nutcracker, were used. When closed, molten iron was ladled into a hole at the top to fill the round internal space. Dipping into a tub of cold water helped to harden it. When cool, the mould was opened and long tongs used to remove the ball which was then 'fettled', smoothed with a file. One imperial ton of iron could make 724 cannon balls.

Horses and carts would have been needed to carry the ordnance to Bridgnorth, Shrewsbury and Ludlow or the pig iron to nearby blacksmiths or forges for making items such as firebacks, horseshoes, tools, nails, chains, pots, kettles, etc.

A cannon ball was found in the garden of John and Jill Perks who renovated what is known as Jacks Barn, buildings on the other side of the lane from Bank Farm. Its diameter was 7.15cm diameter and it weighed 1.4kg. A central ridge line was visible from when it was moulded and the end from the mold had been 'fettled' off.

Cannon balls were made in different sizes 4, 6, 9, 12, 18, 24, 32 and 42 pounds. (1.8, 2.7, 3. 4.5, 5.4, 8.1, 10.8, 14.4 and 18.9 kg.) As the Jacks Barn ball weighed less than the smallest shot. It is possible the reduction was due to rusting whilst in the ground for a few hundred years..

As the Civil War ended in 1651 and the Parliamentary Army did not have the need for as many cannon in peace time, demand dropped and the Bouldon iron masters would have had to diversify. In the early decades of the industry, large, hand-operated 'bellows' were used to blow air in to the furnace to raise the temperature - a back-breaking job. When it was discovered that a water wheel could be used to operate bellows is unknown but once the investment was made, these new bellows greatly increased the temperature in the furnace. These new water-powered furnaces were reported to produce several hundred tons of better-quality iron a year. Needing a regular supply of water, they had to be beside a stream or river or water needed to be diverted to the water wheel, either undershot, turning the wheel from below, or overshot, turning the wheel from above. The wheel was also used to power the drill, a heavy wooden hammer and bellows, being close to a source of running water was essential. This would mean that the new furnace would not have been at the quarry site.

Further up Pye Brook, a stone weir was constructed (SO549846) with a sluice gate on the top. The date of its construction is unknown. When the sluice gate was closed the water-level rose. To provide water power to the furnace, a man-made channel known as a 'leat' or 'mill-race' was dug along the line of the contour, north and then west past what is now Bank Farm. The leat was lined with limestone slabs and the gaps filled with clay to reduce water loss. A second channel ran down the slope from the upper leat, round a 'dog-leg' to run parallel. Both still be seen today, rather overgrown, running alongside the track from Cold Weston into Bouldon and along three boundaries of Bank Farm and the northern boundary of Pool Cottage.

According to the Victoria County History, Bouldon had eight houses which paid hearth tax in 1672, four having more than one hearth. (Hearth Tax 1672, (Salop. Arch. Soc. 1949), p.178) Whether the furnace was taxed was not mentioned but it is possible the furnace manager lived in a house nearby.

During the 17th century, before Bank Farm and its outbuildings were constructed, it is possible that the flat land was the site of a new blast furnace. The leat would have provided water to drive a water-wheel which could have operated a bigger set of bellows. Although there is no evidence of a wheel on the present building, the leat runs almost 2m deep along the south side of the building. There is evidence that the leat continued west over the track to buildings on the opposite side (now renovated and called Jack's Barn) and then along the contour line to provide water power for what is now Bouldon Farm. An underground channel coming Jack's Barn joins the leat from Bouldon Mill which then drains down to rejoin Pye Brook. ('Mill Race' on O.S. Map 1/2,500, Salop. LXV. 13 (1884 edn.))

There is also a large stone-lined three-sided chamber cut into the southern bank of Pye Brook, immediately below Bank Farm (SO547850). Currently heavily overgrown, removal of fallen down trees and undergrowth is needed before excavation and metal detection can provide archaeological evidence.

A stone built charcoal blast furnace at Duddon Bridge, Lancashire, was 6.09m square and 6.4m high. When the Bouldon blast furnace was constructed is unknown but it would have been water-powered. In which case there would have been a water wheel. Overshot waterwheels, like that at Bouldon Mill, were powered by stored water being released on top of the wheel to turn it round. Undershot wheels were turned by the force of moving water at the base. The presence of deep stone-lined trench, the continuing of the leat, runs along the south side of Bank Farm and could well have powered an undershot wheel in an earlier building.

The Victoria County History reported a water mill on Bouldon's manorial estate in 1611 held by the Crumpe family but its exact location was unspecified. (TNA C 66/1921, m. 6) It would have been a wooden structure with a horizontal waterwheel beside the brook. In 1733 it was described as a muncorn mill which ground wheat and rye and is thought this was on the site of present day Bouldon Mill, just below Pool Cottage. (TNA C 103/164, pt. 1, surv. f. 40;IR 29/29/161; IR 30/29/161, field 430) The overshot waterwheel was below a man-made pool constructed to the eastern side of the building, with a sluice gate to release the water when operations started. The pool was supplied with water brought along the lower leat, which was either covered with stone slabs when the slag heap began to encroach or was blocked by a landslip.

Trinder included a description of a blast furnace of this period as

> ...a masonry stack, supported on at least two sides by arches, one accommodating the tuyere pipe, through which air was conveyed from water-powered bellows to the interior, the other the forehearth, from which were tapped molten iron and molten slag, the waste product of the process. Furnaces were usually built into banks or were approached by earthen ramps to facilitate the charging of iron ore, limestone, used as flux, and the fuel at this stage, charcoal. Other components of a furnace complex might include barns for storage or ore and charcoal, and reservoirs where water could be stored to enable the bellows to work throughout a campaign of up to eight months duration. Slag would be piled up around a furnace site, and might be utilised locally as road metal. Furnaces needed to be within convenient distance of iron ore, which in Shropshire came from seams on the coalfields. Charcoal was burned from cordwood, grown as a crop. And might be drawn from an extensive hinterland. (Trinder, op.cit. p.13)

One such water-powered blast furnace was in operation between Chorley and Cleobury, south of Bridgnorth. The Highley village website has two sketches of the furnace which can be seen in the illustrations.

The Alveley website included the same sketch and the following description of how it worked.

> The furnace contents were ignited and a temperature of over 1000°C was reached, helped by regular blasts of air from bellows. The bellows were worked by a water wheel. The typical furnace was set close to a source of ironstone in a well wooded area on a stream. Metallic iron was released from the ore and at once melted; the molten metal could then be run out of the furnace into moulds. This produced cast iron. This form of iron could be cast directly into objects such as pots, pans, kettles, fire grates and backs or smoothing irons. These found a ready market through retail

ironmongers. However, cast iron is a brittle material and much was converted into the more resilient wrought iron. The iron was cast into bars called pigs and passed onto the iron forge. Here they were remelted, perhaps after blending with other types of iron and refined to reduce their carbon content. They were then drawn out under the forge hammer to give bars. These could be sold directly to iron merchants, or passed to slitting mills. At these mills the bars would be rolled and cut into thin rods, particularly suitable for nail making. (http://www.alveleyhistoricalsociety.com/uploads/2/3/2/3/23232952/hampton_loade_furnace_-_poyner.pdf)

The other possibility is that it was a 'shot tower'. A wooden platform from the furnace to the top of a shot tower allowed workmen to carry containers of molten iron to be emptied into an opening. As it fell, the dribbles cooled and solidified into ball-like shapes and landed in a bed of sand or a large iron bowl of water at the bottom of the tower. These balls were then collected and checked for roundness. If mis-shaped, they were filed down or placed in a barrel with other shot to be rolled around to smooth off any imperfections. These would have been sold as shot fired by musketeers. Whether lead shot was manufactured in Bouldon is undocumented.

It is possible that it was fed with water from another leat that had been constructed at a right angle from the upper leat, down the slope in a 'dog-leg' to take the water close to the top of the 'cut'. As in the above diagram, an open wooden channel, supported on wooden stilts, could have taken the water to the top of a water wheel. It is possible that this second leat was constructed at a later date to feed into a pool above Bouldon Mill, where a sluice gate allowed the stored water to fall onto the mill's waterwheel that can still be seen on the south side of the building.

With increased demand for charcoal, much of the available woodland was cut down so that by the end of the 17th century, charcoal was in short supply. By 1702, the furnaces at Abdon, Ifton Rhys, Tislop and Wombridge had ceased operations but new furnaces were erected at Charlcotte (SO 639861) and Kemberton (SJ 744045). (Trinder, Barrie 'The Industrial Archaeology of Shropshire,' Logaston Press, 2016, pp.13-14)

In the late 17th century, Staffordshire-born Abraham Darby worked as an apprentice in the manufacture of brass mills for grinding malt. He would have observed that malt manufacturers had begun to use coke to fire their malting ovens to reduce the sulphur from coal spoiling the taste of ale. Coke was originally manufactured by burning coal in heaps on the ground in such a way that only the outer layer burned, leaving the interior of the pile in a carbonized state. Once cooled, the loose coke could be poured into a furnace with the iron ore and limestone and burnt. Using water-powered bellows, much higher temperatures could be generated.

In 1709, Darby developed a method of producing pig iron in a blast furnace fuelled by coke rather than charcoal, a major step forward in the production of iron as a raw material for the Industrial Revolution. However, rather than purchase this new technology, the owner of the Bouldon furnace continued to use charcoal. Trevor Rowley showed that the cost of charcoal rose 55% over 39 years, from £1.16.0d. in 1736 to £2.16,0d. in 1775. (Trinder, op.cit. p. 30)

According to the Victoria County History, Bouldon furnace produced iron for about 150 years. From 1696 to 1702 or later, the furnace's lessee was William Hall and from 1718 it was Thomas Read. By 1721, Sir Edward Blount (d. 1758) was the tenant, by 1761 Sir Edward Blount, a son or brother (d. 1765), and by 1793 Sir Walter Blount. (Shropshire Archives, Quarter Sessions Records, parcel 281, register of papists' deeds 1717–88, pp. 16, 202, 249; ibid. ff. pp.184–5; S.P.L., MS. 6865, p. 96)

By the early 18th century, the furnace produced mainly pig iron. Annual output at Walker's Bouldon furnace was estimated at 400 tons in 1717 and the total from his other furnaces that year were 2,650 tons. (Transactions of the

Newcomen Society, vol. ix. p.22; Goodman, op.cit. p.275) Richard Amies, who operated Bouldon furnace in the 1720s was also a timber dealer. (Shropshire Archives, quarter session records, parcel 281, reg. of papists' deeds 1717–88, f. 185 and p. 202; 5460/2/3/15–16)

'Piggs' from Bouldon were sold to Mr Charles Lloyd of Dolobran Forge, a few hundred yards down the Vyrnwy from Mathrafal on the Lloyd side of the river. (Lloyd, Humphrey, Quaker Lloyds in the Industrial Revolution 1660 – 1860, Routledge, 1975, reprinted 2006) Ken Goodman reported that its pig iron was supplied to forges as far away as Pool Quay in Guilsfield, Montgomeryshire, Ffridd Mathrafal, Park Mathrafal, both in Llangynyw, Montgomeryshire, and probably to the Blounts' forges at Cleobury Mortimer. (Goodman, 'Hammerman's Hill', op.cit, p.107) Trinder reported that it also supplied pig iron to forges in Cleobury Mortimer and Coalbrookdale Upper Forge and Rowley reported iron from Bouldon, Chalcotte and Bringewood being treated in local forges such as those at Wrickton, Prescott and Cleobury Mortimer, but a considerable quantity was carried by cart to Bridgnorth and thence by barge to Bewdley. (Trinder, op.cit; Rowley, op.cit. p.219)

The Bouldon furnace remained in the Blount family's possession apart from a short period when it was owned and worked by Edward Knight. Richard Knight of Madeley purchased the Bringewood furnace and expanded the business by constructing a forge and slitting mill. Although Bouldon pig iron was sent to Bringewood, demand was so great that the Knights erected another charcoal iron furnace at Charlcotte. Richard and his brother Edward ran the business until it was taken over by Richard's son, also called Richard Knight. Charlcotte had ceased production in the 1780s and Bouldon furnace was sold in 1795. (Trinder, op.cit. p.30)

It is worth mentioning that when Richard Payne Knight, inherited the family's iron business, he used some of the profits to build Downton Castle. There are accounts dated July and August 1774 where he charged five shillings for riding over to Bouldon Quarry to arrange the delivery of stone for the castle's window sills, portico and staircase. Also, fifteen shillings was paid to 'Aingels the Wheelwright's bill for repairs to the Ston Carriage at Bowldon and finding Timber.' Fifteen shillings and two pence was paid to 'Mr Jones the Black Smith's Bill for repairing the carriage' and one shilling and six pence was paid 'for Ale at Didlebury for the waggoners and Assistants when they Brought the first great Ston.' (Herefordshire Archives T74/413,414)

Trinder reported that the Bouldon furnace was one of the principal ironworks in the county but did not state when. In both 1737 and 1795 it was alleged to produce the finest pig iron in England. In 1795 the lord offered it to let as a going concern, and cordwood was still being supplied in 1798 but no later reference has come to light. (Salopian Journal. 5 Aug. 1795; Birmingham University Library, Mytton Papers, Vol. iv. 753.fn. 87)

Romley Wright, employed by the Ordnance Trigonometrical Survey of England, in a paper read to the Geological Society on 18 December 1833 on the Geology of Brown Clee Hill, reported that at the site of the Bouldon furnace, 'Immense heaps of refuse are yet seen, from which the metal has been but very partially extracted.' (Wright, Romley, 'Notes on the Geology of the Brown Clee Hill, in the County of Salop', Transactions of the Geological Society, 18 December 1833)

When the ironworks ceased operations at the end of the 1790s, masonry and a large slag heap were recorded next to the mill. (Shropshire Archives 5403/7/7/2a; Shropshire Magazine Feb. 1966, 29, fn. 88) The building was reported to have been converted into a paper mill near to Furnace field and Furnace wood. (Rowley, 'S. Salop. Landscape', 178–9; TNA, IR 29/29/161; IR 30/29/161, fields 449, 452) Peter Medlicott was recorded as making paper at Bouldon by 1803, when his partnership with Henry Proctor, Thomas Green, and Thomas Lawley was dissolved, and still in 1816. By 1832 the paper mill was run by Simon Cox, (T.S.A.S. liii. 147, 149, 157) who still held it in 1842 as part of the later Bank Farm; it was attached to the farmhouse on a leat from Pye brook. (TNA, IR 29/29/161; IR 30/29/161, field 443) In 1841 the paper makers were William Cox and William Baker. (TNA, HO 107/912/22, ff. 5v., 6) Production had ceased by 1851. (TNA, HO 107/1982, ff. 541–542v)

As well as the quarry in Furnance Wood, there were several other quarries in Bouldon which supplied limestone and sandstone for building purposes. (Salopian Jnl. 5 Aug. 1795; O.S. Map 1", sheet LXI. SW. (1833 edn.); O.S. Map 1/2,500, Salop. LXV. 13 (1884 edn.); TNA, IR 29/29/161; IR 30/29/161, field 437) In 1774 Bouldon supplied stone for building Downton Castle (Herefordshire) (H.W.R.O.(H.), T 74/414, acct.) and in the early 19th century it was one of the best local sources of sandstone flags. (Shropshire Archives 4367/Ch/1, s.a. 1818–19; R. I. Murchison, The Silurian System, (1839), i. 179) John Smallman of Bouldon Rock was a builder in 1811. (W.B.R., Q2/1/3) Bouldon was still producing 'considerable quantities' of building stone in 1850. (S. Bagshaw, Dir. Salop. (1851), 538)

Three masons lived there in 1841 (TNA, HO 107/912/22, ff. 5v., 6v) but none in 1851. (TNA, HO 107/1982, ff. 541–542v) In 1841, Bouldon had a shoemaker, a sawyer, a blacksmith, and a wheelwright. (TNA, HO 107/912/22, ff. 5v.–6v)

During Bouldon Furnace's 150 year history it provided iron not just for the Royalist Army but also to local blacksmiths' forges and iron foundries across

Britain. It brought wealth to its managers, provided long-term employment to its workers and helped stimulate an industrial economy in Corvedale. Miners, carters, cartwrights, wheelwrights, carpenters, builders and others would have benefitted from the business generated with the furnace.

Further research is needed in order to tell more of Bouldon Furnace's story, in particular archaeological excavations and access to landowners' leases and agreements, surveyors' maps and account books.

Bibliography

Books and Journals
Bagshaw, S. Directory Salop. (1851), 538
Baldwin, M. 'Ironworking in Cleobury Mortimer', Cleobury Chronicles, 3 (1994)
Bayliss, D,G. 'The Effect of Bringewood Forge and Furnace on the landscape of part of northern Herefordshire to the end of the seventeenth century', Transactions of the Woolhope Naturalists Field Club, 45/3 (1987)
Bull, Stephens, The Furie of the Ordnance Artillery in the English Civil Wars, Boydell Press, 2008, p.75
Calendar of State Papers, Domestic, 1641–3, p.488; 1644, p.108
Cokayne, George Edward, The Complete Peerage of England, Vol. ii. pp. 202–3
Goodman, Ken, 'Hammerman's Hill: Land, People, and Industries of Titterstone Clee Area 16th to 18th Centuries.' (Keele University, Ph.D. thesis, 1978), p.106
Hayman, Richard , The Shropshire wrought Iron industry between c.1600 and 1900: A study of technological change, University of Birmingham, July, 2003
Hearth Tax 1672, (Salop. Arch. Soc. 1949), p.178
Lloyd, Humphrey, Quaker Lloyds in the Industrial Revolution 1660 – 1860, Routledge, 1975, reprinted 2006)

Murchison, R. I. The Silurian System (1839), i. 179
Roberts, John, The Compleat Cannoniere, London, 1652, pp.24-5
Rotherham, Ian D. Jones, Melvyn and Handley, Christine (eds.), Working and Walking in the Footsteps of Ghosts. Vol. 1: The Wooded Landscape, Wildtrack Publishing, 2003/2012
Rowley, R.T. 'History of South Salop. Landscape', Oxford University, B.Litt. thesis, 1967, p.170
Rowley, Trevor, The Shropshire Landscape, Hodder and Stoughton, 1972
Rowley, Trevor, The Landscape of the Welsh Marches, Michael Joseph, 1986, p.228
Salopian Journal, 5 Aug. 1795
Southcliffe Ashton, Thomas, Iron and Steel in the Industrial Revolution, Manchester University Press, (1953) and 1951, p.18
Transactions of the Newcomen Society, vol. ix. p.22
Transactions of the Shropshire Archaeological Society, 4th ser. Vol. xi. p.109
Transactions of the Shropshire Archaeological Society, Vol. liii. pp.147, 149, 157
Trinder, Barrie, The Industrial Archaeology of Shropshire, Phillimore, 1996, p.30
Trinder, Barrie 'The Industrial Archaeology of Shropshire,' Logaston Press, 2016, pp.13-14
Victoria Country History, Salop. Vol. i. p.473
Wright, Romley, 'Notes on the Geology of the Brown Clee Hill, in the County of Salop', Transactions of the Geological Society, 18 December 1833

Websites
https://www.british-history.ac.uk/vch/salop/vol10/pp147-151
https://www.british-history.ac.uk/vch/salop/vol10/pp151-167
http://www.highley.org.uk/page28.html
https://en.wikipedia.org/wiki/Bringewood_Ironworks
www.ehs.org.uk/dotAsset/6a439da7-0919-41de-acaf-b37ed608496e.xls

https://www.tf.uni-kiel.de/matwis/amat/iss/kap_a/illustr/ia_2_4.html
https://www.haraldthesmith.com/an-introduction-to-iron-smelting-part-i-theory/
https://en.wikipedia.org/wiki/Mountjoy_Blount,_1st_Earl_of_Newport
https://en.wikipedia.org/wiki/Shropshire_in_the_English_Civil_War
http://www.shropshirehistory.org.uk/html/search/verb/GetRecord/theme:20091104222852
https://www.youtube.com/watch?v=e60CAAAhshI; https://www.youtube.com/watch?v=NNMF6BDL8nk
https://en.wikipedia.org/wiki/Viscount_Carrington
https://www.historyonthenet.com/english-civil-war-weapon
https://www.ebay.co.uk/p/King-and-Country-English-Civil-War-Cannon-PNM036-Pnm36/1139307020
http://www.alveleyhistoricalsociety.com/uploads/2/3/2/3/23232952/hampton_loade_furnace_-_poyner.pdf
http://search.shropshirehistory.org.uk/collections/getrecord/CCS_MSA3623/
http://www.alveleyhistoricalsociety.com/uploads/2/3/2/3/23232952/hampton_loade_furnace_-_poyner.pdf

Shropshire Archives
X5735/2/2/2 Brooke papers
X7381/112/753A, IMG1037
John Roque's 1752 map
Robert Baugh's 1808 map
Greenwood's 1828 map
1842 Tithe and Appointment map
1359, box 20, Thomas Martin to Richard Holman, 26 June 1661 (copy)
Quarter Sessions Records, parcel 281, register of papists' deeds 1717–88, pp. 16, 185, 202, 249; ibid. ff. pp.184–5
S.P.L., MS. 6865, p. 96
5460/2/3/15–16
4367/Ch/1, s.a. 1818–19

'Mill Race' on O.S. Map 1/2,500, Salop. LXV. 13 (1884 edn.
O.S. Map 1", sheet LXI. SW. (1833 edn.)

Birmingham University Library
Mytton Papers, Vol. iv. 753.fn. 87

The National Archives
C 66/1921, m. 6
C 103/164, pt. 1, surv. f. 40
IR 29/29/161
IR 30/29/161, fields 430, 437, 443, 449, 452
HO 107/912/22, ff. 5v.–6v
HO 107/1982, ff. 541–542v

Herefordshire Archives
T 74/414, acct.
T74/413,414

Discover more of Bernard O'Connor's publications:
www.lulu.com/spotlight/coprolite

Visit his website:
www.bernardoconnor.org.uk